Bio-Inspired Algorithms in PID Controller Optimization

Intelligent Signal Processing and Data Analysis

Series Editor
Nilanjan Dey

Department of Information Technology
Techno India College of Technology
Kolkata, India

Proposals for the series should be sent directly to
one of the series editors above, or submitted to:
Chapman & Hall/CRC
Taylor and Francis Group
3 Park Square, Milton Park
Abingdon, OX14 4RN, UK

Bio-Inspired Algorithms in PID Controller Optimization
Jagatheesan Kaliannan, Anand Baskaran,
Nilanjan Dey and Amira S. Ashour

Bio-Inspired Algorithms in PID Controller Optimization

Jagatheesan Kaliannan
Anand Baskaran
Nilanjan Dey
Amira S. Ashour

CRC Press
Taylor & Francis Group
Boca Raton London New York

CRC Press is an imprint of the
Taylor & Francis Group, an **informa** business

CRC Press
Taylor & Francis Group
6000 Broken Sound Parkway NW, Suite 300
Boca Raton, FL 33487-2742

First issued in paperback 2020

ISBN-13: 978-1-138-59816-4 (hbk)
ISBN-13: 978-0-367-60696-1 (pbk)

Library of Congress Cataloging-in-Publication Data

Names: Kaliannan, Jagatheesan, author. | Baskaran, Anand, author. | Dey, Nilanjan, 1984- author. | Ashour, Amira, 1975- author.
Title: Bio-inspired algorithms in PID controller optimization / Jagatheesan Kaliannan, Anand Baskaran, Nilanjan Dey, and Amira S. Ashour.
Description: First edition. | Boca Raton, FL : CRC/Taylor & Francis Group, 2018. | Series: Intelligent signal processing and data analysis | "A CRC title, part of the Taylor & Francis imprint, a member of the Taylor & Francis Group, the academic division of T&F Informa plc." | Includes bibliographical references and index.
Identifiers: LCCN 2018014060| ISBN 9781138598164 (hardback : acid-free paper) | ISBN 9780429486579 (ebook)
Subjects: LCSH: Interconnected electric utility systems--Automation. | Electric power systems--Load dispatching--Mathematics. | Mathematical optimization. | Nature-inspired algorithms. | Cogeneration of electric power and heat. | PID controllers.
Classification: LCC TK1007 .K35 2018 | DDC 621.319/1--dc23
LC record available at https://lccn.loc.gov/2018014060

Visit the Taylor & Francis Web site at
http://www.taylorandfrancis.com

and the CRC Press Web site at
http://www.crcpress.com

Contents

Preface

Nowadays, interconnection of different power-generating systems has increased due to the enormous amount of technical growth, industrial development, and modern technologies to satisfy load demand. The automatic generation control (AGC) in power systems handles the sudden load demand and the delivering of stipulated power with good quality in a sudden and continuously varied load period. Stability of standalone power systems and the power quality are affected during the sudden load disturbance. In order to overcome these issues, proper design of power systems and suitable controller modeling is crucial when nonlinearities and boiler dynamics component effects are incorporated in the system. Load frequency control (LFC) has a substantial role in electric power systems with interconnected areas. Reliable maneuver of the power system necessitates the power balance between the system-associated losses and the total load demand of the power generation. Thus, the LFC is used in the power system to keep the frequency and tie-line power flow of the system within the limit during sudden load disturbance.

The main problem in the interconnected power system is reducing the damping oscillations in the system frequency; thus, the tie-line power flow deviations should be kept within the limit during sudden load demand. When damping oscillations exist in the system response for a long period of time without any adequate controller, it affects the system operation and quality of delivered power supply. To provide good quality power and stable power system operation, extensive research work has been carried out and proposed in the last few decades. Due to the massive development in industries and technology, the load demand value is changeable and cannot be predicted as it varies randomly.

Several efforts are carried out based on effective optimization methods to realize numerous benefits and purposes for a power system's operation control. Researchers have conducted different studies to solve the optimization problems to optimize the power system secondary controller parameters. Differential evolution (DE), particle swarm optimization (PSO), firefly algorithm (FA), genetic algorithm (GA), and ant colony optimization (ACO) are examples of optimization algorithms that can be included to design PID controller parameters for effective operation of a thermal power system. In the power systems, the proportional–integral (PI) and proportional–integral–derivative (PID) controllers are used as secondary controllers. Consequently, this book includes different applications of the optimization techniques to design the PID controller for LFC of single area as well as multi-area interconnected thermal power systems with and without incorporating nonlinearities and boiler dynamics effects.

Jagatheesan Kaliannan, PhD
Anand Baskaran, PhD
Nilanjan Dey, PhD
Amira S. Ashour, PhD

Book Objectives

Single area and multiarea power generating system responses are affected during emergency load disturbance conditions, and the power system has more nonlinear components, such as the governor dead band (GDB) and generation rate constraint (GRC) nonlinearities. In order to get desired performance in the power system, all nonlinear component effects are incorporated during optimization of the controller gain values. Therefore, optimization techniques based on bio-inspired algorithms (BIAs) are implemented to tune PID controller gain values and it is considered as secondary controller during emergency load conditions in the power system. The primary objectives of this book are as follows:

- To propose the clear Simulink® model of the single area and multiarea interconnected thermal power system by considering nonlinearities and boiler dynamics effects in power systems.

- To discuss and propose a bio-inspired algorithm–based optimization technique to tune the gain value of the PID controller in single area as well as multiarea interconnected thermal power systems.

- To evaluate the performance of the proposed BIA's tuned controller by comparing other optimization techniques' optimized controller performance in the same system.

Authors

Jagatheesan Kaliannan, PhD, is currently associated with the Department of Electrical and Electronics Engineering, Paavai Engineering College, Namakkal, India. He received his BE degree in electrical and electronics engineering in 2009 from Hindusthan College of Engineering and Technology, Coimbatore, Tamil Nadu, India, and his ME degree in applied electronics in 2012 from Paavai College of Engineering, Namakkal, Tamil Nadu, India. He completed his PhD in information and communication engineering in 2017 from Anna University Chennai, India. His research interests include optimization techniques, advanced control systems, electrical machines, and power system modeling and control. He has published more than 35 papers in national and international journals and conference proceedings, and more than 5 book chapters in reputed books. He is an associate member of UACEE; member of SCIEI, IACSIT, IAENG, and ISRD; and graduate student member of IEEE.

Anand Baskaran, PhD, received his BE in electrical and electronics engineering in 2001 from Government College of Engineering, Tirunelveli, India; ME in power systems engineering from Annamalai University in 2002; and PhD in electrical engineering from Anna University, Chennai, India, in 2011. Since 2003, he has been with the Department of Electrical and Electronics Engineering, Hindusthan College of Engineering and Technology, Coimbatore, Tamil Nadu, India, where he is currently working as an associate professor. His research interests are power system control, optimization, and application of computational intelligence to power system problems. He has published more than 85 papers in national and international journals and conference proceedings. He is a member of IEEE, SSI, and ISTE.

Nilanjan Dey, PhD, is currently associated with the Department of Information Technology, Techno India College of Technology, Kolkata, West Bengal, India. He holds an honorary position of visiting scientist at Global Biomedical Technologies Inc. (California); and research scientist of laboratory of applied mathematical modeling in human physiology, Territorial Organization of Scientific and Engineering Unions, Bulgaria. He is an associate researcher at Laboratoire RIADI, University of Manouba, Tunisia. He is an associated member of the Wearable Computing Research Lab, University of Reading, London. His research topics include medical imaging, soft computing, data mining, machine learning, rough sets, computer-aided diagnosis, and atherosclerosis. He has published 25 books, and 300 international conference and journal papers. He is the editor in chief of the *International Journal of Ambient Computing and Intelligence* and *International Journal of Rough Sets and Data Analysis*; co-editor in chief of *International Journal of Synthetic Emotions (IJSE)* and *International Journal of Natural Computing Research*; series editor of Advances in Geospatial Technologies and Advances in Ubiquitous Sensing Applications for Healthcare (AUSAH), Elsevier; executive editor of *International Journal of Image Mining (IJIM)*; and associated editor of *IEEE Access* and the *International Journal of Service Science, Management, Engineering and Technology*. He is a life member of IE, UACEE, and ISOC. He is also been the chairman of numerous international conferences, including ITITS 2017, WS4 2017, and INDIA 2017.

Amira S. Ashour, PhD, is an assistant professor and head of the Department of Electronics and Electrical Communications Engineering, Faculty of Engineering, Tanta University, Egypt. She received her masters degree in electrical engineering in 2001 and PhD in smart antenna in 2005 from the Department of Electronics and Electrical Communications Engineering, Faculty of Engineering, Tanta University, Egypt. She was the vice chair of the Computer Science Department, Computers and Information Technology College, Taif University, Saudi Arabia, from 2009 to 2015. She was the vice chair of the Computer Engineering Department, Computers and Information Technology College, Taif University, Saudi Arabia, for one year in 2015. Her research topics of interest include smart antennas, direction of arrival estimation, target tracking, image processing, medical imaging, machine learning, soft computing, and image analysis. She has been published in 7 books and 105 international conference and journal papers. Ashour is an editor in chief for the *International Journal of Synthetic Emotions (IJSE)*. She is coeditor of the book series Advances in Ubiquitous Sensing Applications for Healthcare (AUSAH). She is an associate editor for the *International Journal of Rough Sets and Data Analysis*, as well as the *International Journal of Ambient Computing and Intelligence*. She is an editorial board member of the *International Journal of Image Mining (IJIM)*.

CHAPTER 1

Introduction

CONTINUING DEVELOPMENT IN TECHNOLOGY has raised the dependence on electrical power availability. Commercial power resources enable the modern world to operate even with all the required demands. The electric power system has a significant role in several applications including the storage, transfer, and use of electric power. The electric power system is an electrical network of components organized to convert one form of energy into a useful form of electrical power. The operation and performance of electrical equipment is mainly based on the quality of power supply. With the emergence of sophisticated technology, intelligent technology demands power that is free of disturbance and interruption.

The mismatch between power generation and load disturbance affects generating voltage and frequency of the standalone system. In order to overcome this concern, a power-generating unit becomes an essential part of regulating the frequency of the system, tie-line power flow between connected areas, voltage value, and load flow conditions within the desired value. Nonetheless, nowadays, the electric energy demand is rapidly increasing due to the enormous development in technology. Consequently, large-scale power systems are created to balance energy demand with generation. When load demand is increased in any one of the power systems, the remaining connected areas share the power between them to maintain the system in stable condition. The size and complexity of systems are increasing with large interconnections of control areas. The complexity of systems is reduced with the help of recently developed modern control theory.

The power balance among the generating power and the total load demand provides good quality power and reliable power supply to all consumers. During the nominal loading conditions, each power-generating unit takes care of its stability and operating point. Whenever sudden load disturbance occurs in any of the generating units, it disturbs the frequency of the system and tie-line power flow within the interconnected power systems. In addition, the tie-line power flow deviations between the interconnected power plant and damping oscillations that occur in the system response affect the stability of the power system. In order to guarantee power system stability, a speed governor can be used to act as a primary control loop; in addition, a secondary controller can also be introduced to retain the system parameters within the quantified value. The frequency regulation and active power control is called load frequency control (LFC). Furthermore, the voltage control and reactive power are referred to as automatic voltage regulation (AVR). The main aim of LFC is to preserve the frequency with a constant value even with the continuous change of the active power demand, which is related to the loads, and it regulates the tie-line power exchange between the interconnected areas.

The electrical grid of interconnected networks is used to deliver the electrical supply from the generating unit to power consumers. The grid contains power-generating stations to generate the electrical power supply. The high voltage transmission line carries the generated power from the sources to load centers, where the distribution centers connect individual customers and where the power-generating stations are positioned near the fuel source.

1.1 LOAD FREQUENCY CONTROL AND AUTOMATIC GENERATION CONTROL

The automatic generation control (AGC) as well as the LFC play the foremost role in the power system process and the control of any type of power plant unit. For generating and power delivery, suitable analysis of the power supply quality and regularity become essential during emergency situations. In this situation, power-generating units are interconnected via tie-lines to obtain good quality of power supply. The interconnected power system comprises hydro, nuclear, wind, solar, gas, and thermal power plants. Thus, the system response yields damping oscillations in the frequency and tie-line power flow deviations during the load demand condition. To deliver good quality power supply to consumers, the secondary controller gain values should change uniformly. During the higher-load demand condition, controller gain values change maximally and this repeating process maintains the quality of generated power by keeping error values equal to zero.

Typically, the PID controller consists of three basic terms: proportional controller (P), integral controller (I), and derivative controller (D). The proportional controller reduces the peak overshoot in the system responses, the integral controller reduces the steady-state error to zero, and the stability of the system is increased by using the derivative controller. The input of PID controller is the area control error (ACE) and the output of the controller is the control signal (delP$_{ref}$). The output of controller is given into the power system as a reference signal.

The PID controller design in the literature review included several methods for optimization of controller gain values with different cost functions. Such functions include the integral square error (ISE), integral time square error (ITSE), integral absolute error (IAE), and integral time absolute error (ITAE) cost functions. In this book, the PID controller is considered a feedback controller, which gives the appropriate control signal for controlling the power plant during abnormal conditions. The value of the control signal generated by the controller for each area is given in the following expression:

$$\Delta P_{ref} = u(t) = -K_p.ACE - \frac{K_i}{T_i}\int ACE - K_d T_d \frac{d}{dt}ACE \qquad (1.1)$$

ΔP_{ref}, or $u(t)$, represents the control signal generated by the controller. In addition, K_p, K_i, and K_d represent the proportional, integral, and derivative controller gain values, respectively. ACE indicates the area control error. Based on the IEEE standards, ACE is defined as linear combinations of change in frequency and tie-line power flow deviations between connected areas. So, proper control of the ACE within the tolerance value keeps the system parameters within the nominal value during sudden load disturbance. The deviations in frequency and tie-line power flow deviations will be zero when the value of ACE is zero. The value of ACE is given in the following expression:

$$ACE_i = \Delta F_i.B_i + \Delta P_{tiei-j} \qquad (1.2)$$

where B represents frequency bias constant. The subscripts i, j indicate area, that is, $i, j = 1$, 2, and 3. In these conditions, optimization techniques based on bio-inspired algorithms (BIAs) are implemented to tune controller gain values for providing adequate and suitable control signals by optimizing controller gain values. Recently, different controllers have been designed based on bio-inspired algorithms optimization techniques to tune the controller gain values for powerful implementation of the LFC and AGC of the power system.

1.2 BIO-INSPIRED OPTIMIZATION ALGORITHMS

Bio-inspired algorithm–based optimization techniques are implemented based on the behavior of natural living beings. These algorithms are encouraged by biological mechanisms. The natural phenomena are observed by grouping these algorithms, which are used to solve issues related the mathematics. The computational algorithm–based techniques are designed and optimized based on the inputs from the natural behavior of biological systems, such as the bee or ant colonies.

The implementation of suitable secondary controllers is more essential for getting superior, controlled dynamic performance in the interconnected power system during sudden load demand situations. In this book, a PID controller is considered a secondary controller. The major aim of implementing a secondary PID controller is to regulate the power system response by eliminating or reducing the time-domain specification parameters, such as steady-state error, minimum damping oscillations, peak frequency, and the tie-power flow during sudden load demand in any interconnected power system.

The tuning of optimal gain values of proportional, integral, and derivative gain values, K_p, K_i, and K_d, respectively, are crucial. First, find the integral controller gain value by keeping K_i constant, and tuning the K_p gain value. Similarly, by keeping K_i and K_p constant, one can then tune the derivative controller gain value K_d. This type of tuning method is called a trial-and-error method. The system response yields more damping oscillations in the frequency and tie-line power flow deviations during sudden load demand period. To deliver a good quality control signal to the power system, the controller gain values should change equally. During the higher-error value condition, the controller gain value should change maximally and this process repeats until the error value is zero. In this condition, BIA-based optimization techniques are implemented to tune controller gain values to provide adequate and suitable control signals by optimizing controller gain values.

In a system, groups of components are connected together to perform some specific operation or function. The output of a system is controlled by varying input quantity of the system is called a control system. The output of the system is the controlled variable and input of the system is the command signal. Generally, the system can be classified into two types, namely, open-loop system and closed-loop system. In the open-loop system, the system output is affected by the input signal. However, in the case of a closed-loop system, the output of the system is determined by the input of the system. For regulating the output of the system, a controller device is introduced to the system. The controller modifies the error signal by generating an appropriate control signal to the system. The basic control signals that are commonly available in the two-position analog controller are proportional, integral, derivative, proportional–derivative, and proportional–integral–derivative control actions. In this research work, the PID controller is introduced as a secondary controller. The parameters of the PID controller are optimized by using BIA-based optimization techniques for superior system performance.

1.3 LITERATURE SURVEY

The gain values of the fuzzy PID controller are optimized using teaching-learning based optimization (TLBO) that B. K. Sahu et al. [1] proposed for tuning of fuzzy PID controllers for the AGC of a two unequal-area thermal power systems. A hybrid particle swarm optimization–pattern search (hPSO-PS) procedure is applied in the AGC for optimizing the fuzzy PI controller gain values in a multiarea interconnected power system [2]. R. K. Sahu et al. [3] obtained the PI controller gain values using the minority charge carrier inspired (MCI) algorithm in an AGC of an interconnected hydrothermal power system. The bat-inspired algorithm has been applied to optimize the PI controller gain values for a multiarea interconnected power-generating system to solve LFC issues such as frequency deviations and tie-line power flow deviations and stability of the power system [4]. A fractional order PID (FOPID) controller has been designed for the LFC of an interconnected power generating system, where the controller gain values were optimized using a multiobjective optimization method [5].

Dash et al. [6] used the cuckoo search (CS) algorithm to tune two degrees of freedom (2-DOF) controllers for solving the AGC issue (frequency deviations and tie-line power flow deviations within interconnected power systems) in a multiarea interconnected power system with several flexible alternating current transmission systems. A PD-PID cascade controller has been designed for solving the AGC issue in multiarea interconnected thermal power systems by considering the generation rate constraint nonlinearity effect [7]. A hybrid firefly algorithm–pattern search (hFA-PS) method designed controller has been applied for solving the AGC issue in multiarea interconnected power systems by considering the integral time absolute error objective function [8]. Sahu et al. [9] implemented a hybrid local unimodal sampling (LUS) and TLBO technique tuned fuzzy PID controller into the LFC of interconnected multisource power systems.

Shankar and Mukherjee [10] applied the quasi oppositional harmony search technique to optimized classical controller gain values into the LFC of multiarea, multisource power generating-systems. Vukarasu and Chidmbaram [11] used the bacterial foraging optimization (BFO) algorithm to optimize the proportional–double integral (PI^2) controller in the AGC of interconnected power systems under deregulated environments. The fuzzy PID controller parameters are optimized using the firefly algorithm, which have been applied into the AGC of multiarea, multisource power systems with a unified power flow controller (UPFC) and superconducting magnetic energy storage unit by Pradhan et al. [12]. Sharma and Saikia [13] used the grey wolf optimizer algorithm–based classical controller in a multiarea solar thermal–thermal power system as a secondary controller.

Francis and Chidambaram [14] proposed a teaching–learning-based optimization technique for tuning of PI controller gain values and PI+ controller gain values. Shivaie et al. [15] proposed a modified harmony search algorithm–tuned PID controller for the LFC of interconnected nonlinear hydrothermal power systems. The PI/PID controller gain values have been optimized using the firefly algorithm in the LFC of multiarea interconnected thermal power systems. The proposed optimization technique's tuned controller performance has been compared to the genetic algorithm, bacteria foraging optimization technique, differential evolution optimization algorithm, particle swarm optimization technique, and Ziegler-Nichols technique–based controller performance of the investigated power generating system [16]. Prakash and Sinha [17] presented a neuro-fuzzy hybrid intelligent PI controller for the LFC of a four-area interconnected power system. Farhangi et al. [18] applied an emotional learning–based intelligent controller for the LFC of an interconnected power system by considering the generation rate constraint nonlinearity effect. Farook and Raju [19] proposed a hybrid genetic algorithm–firefly algorithm in the AGC of an interconnected three-area deregulated power system. Moreover, a self-adaptive modified bat algorithm has been implemented for optimization of controller gain values in a four-area interconnected power generating system [20].

The AGC of an equal three-area thermal–thermal–hydro power system has been investigated with different classical controllers [21]. The performance is compared to the fuzzy integral double derivative (IDD) controller. The controller gain values are optimized by implementing the BFO technique. The results established that this technique is superior to existing methods. The imperialist competitive algorithm (ICA) has been implemented for the LFC of a three-area power system with different generating units and a fractional order PID (FOPID) controller [22]. The simulation result proved the superiority of the system performance with ICA-based controller compared to the existing controller.

Some other types of controllers and optimization techniques used in the LFC of power systems include the fuzzy logic controller [23], genetic algorithm (GA) [24], artificial neural network (ANN) [25,26], variable structure control (VSC) [27], Lyapunov technique [28], continuous and discrete mode optimization [29], adaptive controller [30], parameter-plane technique [31], optimal control [32], optimal tracking approach [33], decentralized controller [34], ant colony optimization (ACO) [35–37], bacteria foraging optimization (BFO) [38,39] artificial bee colony (ABC) [40], particle swarm optimization (PSO) technique [41], stochastic particle swarm optimization (SPSO) [42], bacterial foraging optimization algorithm (BFOA) [43], bacterial foraging (BF) technique [44], bat-inspired algorithm [45], beta wavelet neural network (BWNN) [46], and cuckoo search (CS) [47,48].

From the aforementioned studies, it is clear that in recent years several optimization techniques and optimized controller gain values-based controllers are considered for improvement of power system performances during sudden load demands.

CHAPTER 2

Load Frequency Control of Single Area Thermal Power System with Biogeography-Based Optimization Technique

THIS CHAPTER CONSIDERS A load frequency controller for a single area thermal power-generating unit by considering different bio-inspired algorithm (BIA)–based optimization techniques for tuning PID (proportional–integral–derivative) controller performance with 1% step load perturbation (SLP). The PID controller is presented as a secondary controller to control the parameters of the system within the specified value during the sudden load disturbance. The gain values of the implemented controller are tuned by using various bio-inspired algorithms, such as simulated annealing, genetic algorithm, particle swarm optimization, and biogeography-based optimization techniques.

The structure of this chapter is as follows. The first section, "Investigated Thermal Power System," presents the Simulink® model of the investigated single-area thermal power system. Section 2.2, "Controller Design and Objective Function," presents the details of the proposed controller and the necessary cost function for tuning of controller gain values. The subsequent section, "Biogeography-Based Optimization Technique," delivers the proposed optimization technique details, and the "Results and Analysis" section demonstrates the effectiveness of the proposed technique. The "Conclusion" describes the performance of different BIA-based optimization techniques over the proposed optimization technique.

2.1 INVESTIGATED THERMAL POWER SYSTEM

In order to model the investigated power system with the proposed optimization based controller, a MATLAB® Simulink® model of the single-area thermal power system is illustrated in Figure 2.1. A thermal power-generating system comprises a turbine with a reheater, speed governor, generator unit, and PID controller. A 1% SLP is applied into the power for the analysis of power system with and without any load demand. The nominal parameters of the investigated thermal power system are as follows: T_g, governor time constant = 0.2 s; K_r, steam turbine reheat coefficient = 0.333; T_r, steam turbine reheat time constant = 10 s; T_t, steam turbine time constant = 0.3 s; K_p, power system constant = 120 Hz pu^{-1}MW; T_p, power system time constant = 20 s; and R, speed regulation = 2.4 Hz pu^{-1}MW.

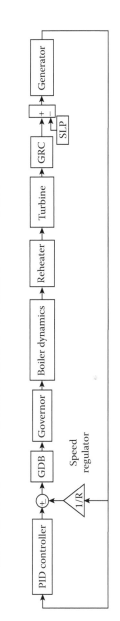

FIGURE 2.1   Simulink model of single-area thermal power system.

The area control error (ACE), which is the input of the investigated power system, is specified as follows:

$$ACE = \Delta f \qquad (2.1)$$

where Δf is the change in frequency deviation of the investigated power system. Figure 2.1 illustrates the MATLAB Simulink of a single-area thermal power system model.

The secondary PID controller is equipped for regulating power system parameters during a sudden load demand condition. The thermal power system incorporates the governor, reheater, turbine, speed regulator, and generator units. When a load disturbance occurs in the open-loop power system, system parameters are affected and yield more damping oscillations with steady error. In order to overcome this issue, secondary controllers are introduced to regulate power system parameters.

2.2 CONTROLLER DESIGN AND OBJECTIVE FUNCTION

Generally, the PID controller consists of a proportional controller, integral controller, and derivative controller. The proportional controller steadies the gain, yet it yields a steady-state error. The integral controller eliminates the steady-state error and the derivative controller reduces change of error rate. The structure of the PID controller under concern is shown in Figure 2.2.

The PID controller input is the ACE and the output is the control signal (U), where the output is given as follows:

$$\Delta P_{ref} = U = -K_p \cdot ACE - \frac{K_i}{T_i} \int ACE - K_d T_d \frac{d}{dt} ACE \qquad (2.2)$$

where $\Delta P_{ref} = \underline{U}$ is the output control signal, K_i is the integral controller gain, K_p is the proportional controller gain, K_d is the derivative controller gain, T_i is the integral time constant, T_d is the derivative time constant, and ACE is the area control error.

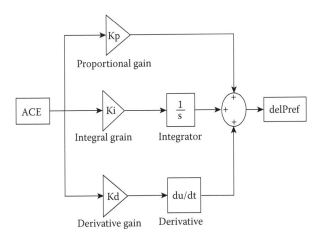

FIGURE 2.2   Arrangement of the PID controller.

The objective function is defined based on the preferred specification and constraint for design of modern BIA-tuned controllers. In this work, the integral time absolute error objective function is considered, which is stated by

$$J = \int_0^T t.\left|\Delta f\right| dt \qquad (2.3)$$

where t is the simulation time. The controller gain values of integral (K_i), proportional (K_p), and derivative (K_d) are optimized by using any of the BIAs.

2.3 BIOGEOGRAPHY-BASED OPTIMIZATION TECHNIQUE

In this research work, the biogeography-based optimization (BBO) technique is proposed for the optimization of gain values in load frequency control (LFC) of a single-area thermal power system. The BBO technique is a population-based optimization technique [49,50]. The human reproduction process is not involved in this optimization algorithm. The BBO optimization technique differs from other population-based optimization techniques. For example, the ant colony optimization technique generates a new solution for each iteration during the optimization process [51,52], whereas the BBO technique keeps the solution set from one iteration to the next iteration. The BBO strategies are common with the particle swarm [53] and differential evolution techniques [54,55]. In these technique-based algorithms, tuned gain values are retained one iteration to the next. Each generated solution has the capability to learn from its neighbor and can adjust itself for tuning the technique process. Particle swarm optimization refers to the corresponding solution as a point in space. It indicates the change over time of each solution of the velocity vector. The selection of differential evolution algorithm parameters can have a great role in performance of the optimization. The choice and selection of DE optimization parameters yield better performance, so it has been the topic of a great deal research.

The biogeography-based optimization technique is an evolutionary algorithm (EA). The optimization technique is implemented for tuning of controller gain values that optimize a cost function by stochastically and iteratively improving the solutions to a given measure of cost function or superiority. The BBO algorithm steps are as follows [49]:

Step 1—Initialize the parameters. In this step, the driving methods are indicated for mapping issues including the suitability index variable (SIV) and habitats. In addition, the maximum species count S_{max}, maximum mutation rate m_{max}, maximum migration rate, and an elitism parameter are initialized.

Step 2—Initialize a random set of habitats to represent implementation of the $\mathcal{I}$ operator.

Step 3—Each habitat maps the Habitat Suitability Index (HSI) to the number of species S, emigration rate μ, and immigration rate λ.

Step 4—Probabilistically use the emigration rate and immigration rate to change each nonelite habitat and recompute every HSI.

Step 5—For every habitat, update its probability of species count by applying initialization of a random set of habitats. Afterward, based on the probability, mutate each nonelite habitat and recompute HSI.

Step 6—Go to step 3 for the next iteration. After a predetermined number of iterations, this loop can be terminated. This is the implementation of the T operator.

2.4 RESULTS AND ANALYSIS

Different BIA-based optimization techniques tuned to PID controller performance are compared by considering 1% step load perturbation in the investigated thermal power system under MATLAB Simulink environment for the simulation period time of 120 s. The effectiveness of the proposed tuning algorithm is analyzed by comparing the frequency deviations and area control error deviations with other techniques for controller tuning performance. By using this proposed algorithm with the ITAE cost function, the PID controller gain values are optimized. The optimal PID controller gain values of different BIA-tuned values are tabulated in Table 2.1. The table reports the PID controller gain values of simulated annealing, genetic algorithm, particle swarm optimization, and BBO technique-optimized controller gain values by considering 120 s as a simulation time period. It is clearly evident that the proposed BBO optimization technique yielded minimum performance index compared to other optimization techniques. The comparisons of simulation results based on

TABLE 2.1   Optimal Gain Values of PID Controllers Obtained by Using Different BIAs

BIA Algorithm	K_P	K_I	K_D	Performance Index
Simulated annealing (SA)	0.93245254	0.63703068	0.0213324	0.16754546
Genetic algorithm (GA)	0.85542643	0.89133524	0.13190307	0.14811123
Particle swarm optimization (PSO)	0.96310035	0.9929861	0.08711779	0.13001764
Biogeography-based optimization (BBO)	0.99999989	0.99999917	0.00099059	0.12606583

different BIA-based optimization techniques are given for a clear understanding.

2.4.1 Response of System with and without Load Demand

The system response does not yield any damping oscillations and steady-state error values. However, when a load disturbance occurs in a power system, the responses yield more damping oscillations with steady-state error. Also the system takes more time to settle with particular error values. Figure 2.3 shows the open loop performance comparisons of a proposed power system with and without any load demand. The solid line shows the open-loop response of a system without any load demand. The dashed line shows the response of system with 1% SLP in the investigated power system.

Figure 2.3 establishes that the comparisons of open-loop performance of the system response do not yield any oscillations, peak overshoot and undershoot, and steady-state error for zero load demand (peak undershoot = 0 Hz and steady-state error = 0 Hz). However, when load demand exists, the response yields more damping oscillations (peak undershoot = −0.058 Hz and steady-state error = 0.026 Hz). In order to overcome this issue, a controller is introduced to get good quality power.

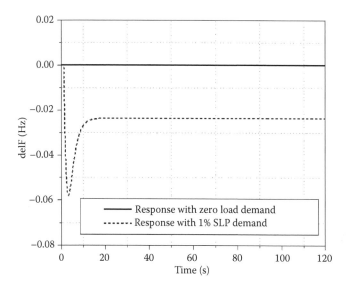

FIGURE 2.3 Dynamic response of system with and without considering load demand.

2.4.2 System Response with Different BIAs Tuned PID Controller

The frequency deviations of PID controller responses to different BIA-based optimization techniques are illustrated in Figure 2.4. It is clearly evident that at up to 10 s all the techniques for controller tuning yielded damping oscillations with error. The BBO PID controller gave the fastest settled response compared to other optimization technique tuned PID controller responses. The settling time of each optimization technique is given in Table 2.2 and comparison responses are shown in Figure 2.5. The results indicate that the PID controller tuned with the BBO technique settled quicker compared to other BIA-based optimization techniques during sudden load demand in the investigated power system.

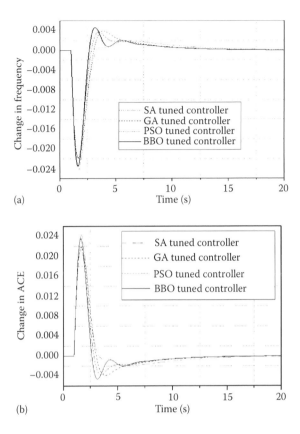

(a)

(b)

FIGURE 2.4 (a) Comparison of changes in frequency for controllers tuned with different BIA optimization techniques. (b) Comparison of changes in area control error for controllers tuned with different BIA optimization techniques.

TABLE 2.2 Numerical Values of Settling Time of PID Controllers Tuned with Different BIAs

Controller	Settling Time in delF (s)	Settling Time in ACE (s)
SA-PID controller	25	27.5
GA-PID controller	22.5	24.8
PSO-PID controller	22	24.7
BBO-PID controller	21.6	24.5

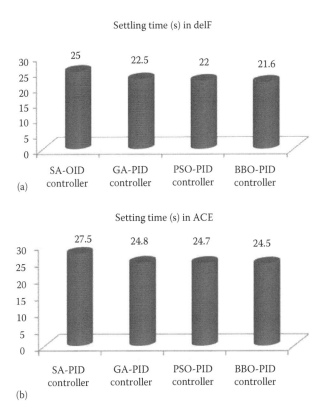

(a)

(b)

FIGURE 2.5 (a) Comparisons of settling time in delF with different BIA-tuned PID controllers. (b) Comparisons of settling time in ACE with different BIA-tuned PID controllers.

2.5 CONCLUSION

The load frequency control of a single-area thermal power system is examined with PID controllers considering 1% step load perturbation. The controller gain values are optimized by implementing different bio-inspired algorithm–based optimization techniques. The simulation result comparisons clearly reveal that the BBO technique tuned controller responses settled faster compared to controllers tuned with other BIAs in the same investigated power system during sudden load demand or emergency conditions.

CHAPTER 3

Automatic Generation Control with Superconducting Magnetic Energy Storage Unit and Ant Colony Optimization-PID Controller in Multiarea Interconnected Thermal Power System

THE ANT COLONY OPTIMIZATION (ACO) technique is implemented for tuning of the PID controller gain values in a multiarea interconnected power system. Once the load disturbance arises in any interconnected power system, it disturbs the system performance from its nominal values. In order to mitigate this issue, a secondary PID controller is implemented. In addition, a superconducting magnetic energy storage (SMES) unit is implemented to improve the system performance. The PID controller gain value is tuned by using the ACO technique.

3.1 SYSTEM UNDER STUDY

The transfer function model of an uncontrolled three-area interconnected thermal power system is given in Figure 3.1 [23,69]. Each thermal power plant has a speed regulator, a governor, turbine with reheater unit, and generator. Each plant has a power rating of 2000 MW. In the single-area power system, a single power-generating unit is connected to a load. During a sudden or emergency load condition, performance and stability of the power system are affected. In this regard, one or more power-generating units are interconnected via a tie-line. During normal loading conditions, it takes care of its own load demand, but when sudden load demand happens in the interconnected power system it shares the power within the interconnected power units to maintain power balance and stability.

All the power plants are interconnected with the help of the tie-lines for interchanging the power among the plants during abnormal condition to keep the system stabile. Increasing the load demand in either plant of the system affects the reactive power and real power of the system. Changes in real power mainly affect the stability of the total interconnected power plant compared to the reactive power changes. In order to enhance the system stability, a proper control signal generated by the controller is fed to the connected plants. The required control signal generated by the PID controller is

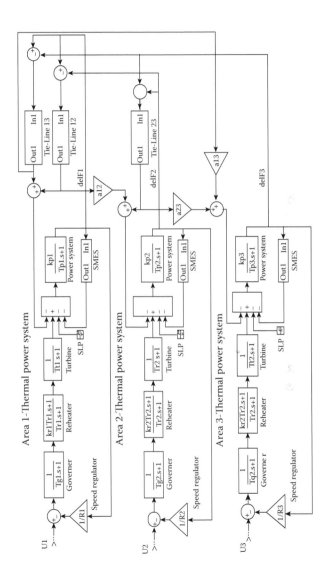

FIGURE 3.1   Open-loop Simulink model of interconnected three-area thermal power system with SMES unit.

$$u_1 = -K_{p1}.ACE_1 - \frac{K_i}{T_i}\int ACE_1 - K_{d1}T_d\frac{d}{dt}ACE_1 \qquad (3.1)$$

$$u_2 = -K_{p2}.ACE_2 - \frac{K_i}{T_i}\int ACE_2 - K_{d2}T_d\frac{d}{dt}ACE_2 \qquad (3.2)$$

$$u_3 = -K_{p3}.ACE_3 - \frac{K_i}{T_i}\int ACE_3 - K_{d3}T_d\frac{d}{dt}ACE_3 \qquad (3.3)$$

where u_1, u_2, and u_3 are the output control signal, K_i is the integral controller gain, K_p is the proportional controller gain, K_d is the derivative controller gain, T_i is the integral time constant, and T_d is the derivative time constant, where i is the size of the power system (i = 1, 2, 3).

The ACE is the linear combination of the system frequency error (delF) and tie-line power flow error (delPtie) between the interconnected power plants. The generated ACE signal in area 1, area 2, and area 3 is given in the following equations:

$$ACE_1 = \Delta f_1.B_1 + \text{del}Ptie_{12} \qquad (3.4)$$

$$ACE_2 = \Delta f_2.B_2 + \text{del}Ptie_{23} \qquad (3.5)$$

$$ACE_3 = \Delta f_3.B_3 + \text{del}Ptie_{31} \qquad (3.6)$$

where ACE is the area control error, B is the frequency bias constant, Δf is the frequency deviation, and ΔP_{tie} is the tie-line power deviation.

3.2 SUPERCONDUCTING MAGNETIC ENERGY STORAGE (SMES) UNIT

The SMES is a device that stores the energy in the form of a magnetic field [56–62]. The basic components of this unit are shown in Figure 3.2. It consists of a superconducting coil, cryogenic system,

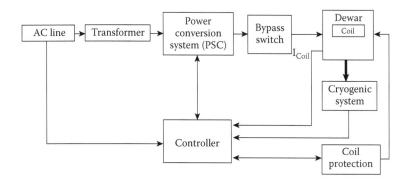

FIGURE 3.2 Components of an SMES unit.

transformer, power conversion system (PCS) with proper control, and protection function arrangement. From these components, the superconducting coil is the core of the SMES unit. It has vacuum vessels and liquid vessels for the cooling arrangement of the coil. A cryogenic system is used for keeping the temperature of the coil within the permissible limits. The power conversion system is used for two purposes, namely, (1) the conversion of electrical energy from DC (direct current) to AC (alternative current) and (2) charging and discharging of the oil. The power system is connected to the PCS with the help of a transformer to reduce the operating voltage to an acceptable level for the PCS.

In the SMES unit, energy is stored in the superconducting coil (E) and the rated power (P) is given by [56–62]

$$E = \frac{1}{2}LI^2 \tag{3.7}$$

$$P = \frac{dE}{dt} = LI\frac{dI}{dt} = VI \tag{3.8}$$

where P is the rated power of the coil, L is the inductance of the coil, V is the voltage across the coil, I is the DC current flowing through the coil, and E is the energy stored in the coil. When the superconducting coil reaches the rated current, the SMES unit is ready for automatic generation control (AGC) or load frequency control (LFC). The ACE is sensed and used to control the voltage of the SMES unit by changing the duty cycle of the chopper. Table 3.1 reports the few noteworthy applications of SMES in LFC/AGC issues of multiarea interconnected power systems.

From the applications of SMES units in interconnected power systems noted in Table 3.1, it is clear that a sudden power surplus in a power system is overcome by deriving power from a large inductor coil. It improves the overall dynamic performance of the system and stability.

3.3 PID CONTROLLER DESIGN

A commonly used industrial PID controller is implemented for solving frequency and tie-line power flow deviations between interconnected power systems. The ant colony optimization technique is implemented for obtaining the gain values of the controller. Arrangement of proposed PID controller is given in Figure 3.3. This PID controller structure consists of three different control terms, namely, the integral, proportional, and derivative terms, where all are connected to solve the AGC problem in the power system. From the literature survey, it is clear that the integral square error (ISE), integral time square error (ITSE), integral absolute error (IAE), and integral time absolute error (ITAE) objective functions are considered and used for the optimization of controller gain values. Here, the ITAE objective function is considered for obtaining the best gain values of K_p, K_i, and K_d. The expression of ITAE cost function is given by

$$J = \int_{0}^{\infty} t \, | \, \{\Delta f_i + \Delta P_{tiei-j}\} \, | \, dt \qquad (3.9)$$

where J is the performance index, t represents the simulation time period, Δf is the frequency deviation, and ΔP_{tie} is the tie-line power deviations $(i, j = 1, 2, 3)$.

TABLE 3.1   Applications of SMES Units in Multiarea Interconnected Power Systems

Size of the Power System	Author/Year	Reference	Remarks
Two areas	Pothiye et al. (2007)	63	AGC problem of a two-area reheat thermal power system is discussed with a fuzzy logic–based PID controller using multiple Tabu search algorithms.
Two areas (multiunit)	Bhatt et al. (2010)	67	Multiarea, multiple unit hydrothermal power system operation is improved by SMES units coordinated by thyristor controlled phase shifters (TCPS) in all areas with an integral controller.
Two areas	Rajesh et al. (2011)	64	Damping oscillations in frequency and tie-line power in a two-area thermal system is discussed with SMES and TCPS coordination.
Two areas (multiunit)	Roy et al. (2014)	65	Dynamic performance of two-area hydropower system with three multiple units in each area is improved by introducing SMES units.
Two areas (multiunit)	Padhan et al. (2014)	66	Multiarea, multiple-unit thermal power system is discussed with fuzzy-based PID controller and it improves system response with less settling time and overshoots.
Two areas	Chaine et al. (2015)	68	The real power surplus in a two-area power system is compensated by deriving the same amount of power from a large indicator coil. The use of an SMES unit in an interconnected two-area power system improves the performance of all interconnected areas and tie-line power flow between the control areas.

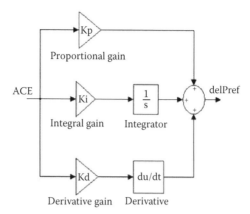

FIGURE 3.3   PID controller.

3.4 ANT COLONY OPTIMIZATION

The natural behavior of real ants inspired many researchers to solve discrete optimization problems. When ants are searching for food, initially all the ants explore in a random manner around their nest. After reaching a food source, the quality and quantity is evaluated. During the return trip, the ant carries some food to the nest and deposits a pheromone chemical trail on the ground. The quality and quantity of the source is determined by the pheromone trial. This is useful for guiding other ants toward the correct food source. The indirect communication among the ants using chemical trails enables the shortest path between food and the nest. These characteristics of real ants are used to find the solution for many optimization problems. The transition probability between towns i and j for the kth ant is as follows:

$$p_{ij}(t) = \frac{\tau_{ij}(t)^{\alpha}(\eta_{ij})^{\beta}}{\displaystyle\sum_{j\in nodes} \tau_{ij}(t)^{\alpha}(\eta_{ij})^{\beta}} \qquad (3.10)$$

where p_{ij} is the probability between the town i η_{ij} and j; τ_{ij} is the pheromone associated with the edge joining cities i and j; α and β

are constants that find the relative time between pheromone and heuristic values on the decision of the ant. The pheromone value versus the heuristic information is

$$\eta_{ij} = \frac{1}{d_{ij}} \tag{3.11}$$

where d_{ij} is the distance between cities i and j. The global apprising rule is realized in the ant system, where all the ants start their tours. The pheromone chemical is deposited and updated on all edges based on

$$\tau_{ij}(t+1) = (1-\rho)\tau_{ij}(t) + \sum_{\substack{k \in colony\ that \\ used\ edge\ (i,j)}} \frac{Q}{L_k} \tag{3.12}$$

where τ_{ij} is the pheromone associated with the edge joining cities i and j; Q is constant; L_k represents the length of the tour performed by the kth ant; and ρ indicates evaporation rate.

PID controller gain value optimization is obtained by using the ACO algorithm and it involves several phases. The first phase is the initialization of simulation parameters, such as the number of ants, pheromone and iteration, and so on after the initial simulation model is run and ants start moving from source to node (nest to food source) by secreting pheromone chemicals into the ground. The concentrations of pheromone are varied based on the quality and quantity of the food. During this tour, the shortest path having higher chemical concentration and the longer path having lesser concentration evaporates after a few iterations. The simulation stops when it reaches the maximum iteration value. The optimal control gain values and results comparisons are given in Table 3.2.

3.5 SIMULATION RESULTS AND DISCUSSION

The dynamic response of the investigated power system (area control error, frequency deviations, and tie-line power flow deviations) in all areas is obtained with help of MATLAB 7.5 (R2007b)

TABLE 3.2   PID Controller Gain Values Optimized by ACO

	PID Controller Gain Values								
	Ki			Kp			Kd		
	Ki1	Ki2	Ki3	Kp1	Kp2	Kp3	Kd1	Kd2	Kd3
Without considering SMES unit	9.6	6.7	5.7	10	9.6	8	1.3	4.3	2.1
Considering SMES unit	0.8	9.6	9.6	6.3	9.6	9.8	8.7	2.3	3.3

in the cases of with and without the SMES unit effect. Table 3.2 reports the gain values of the PID controller that are optimized using the ACO algorithm.

The simulation performance comparison of the AGC multiarea interconnected reheat thermal power system with and without considering an SMES unit in all the areas is shown in Figure 3.4. In Figure 3.4, the dashed line shows the response system without considering an SMES unit and the solid line shows the response system considering an SMES unit in the investigated power system. The PID controller parameters are tuned by using the ACO technique.

Figure 3.4a through c demonstrates the frequency deviations comparisons of the system in area 1, area 2, and area 3 with an ACO-based PID controller with and without considering an SMES unit effect in all areas. The figure shows that system response including SMES unit response settled quickly compared to system response without considering the energy storage unit ($f_1 = 10.1$ s > 10 s, $f_2 = 19$ s > 16 s, $f_3 = 16.5$ s > 10.1 s). In addition, the system frequency response yields minimum damping oscillations with lesser peak undershoot compared to the response without an SMES unit. Figure 3.4d through f represents performance of tie-line power flow deviations between connected power systems with and without considering an SMES unit in all areas and 1% SLP in area 1.

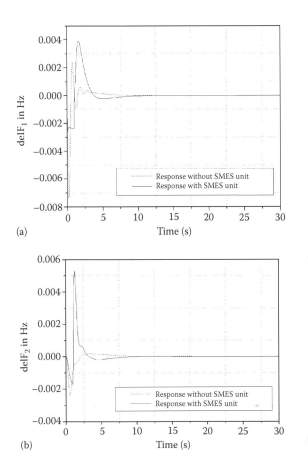

FIGURE 3.4 (a) Comparisons of frequency deviations in area 1. (b) Comparisons of frequency deviations in area 2. (*Continued*)

Figure 3.4d through f illustrates that the tie-line power flow deviations between connected power systems are effectively regulated with good settling time with minimum damping oscillations compared to tie-line power deviations without considering an SMES unit in all areas, where settling times of tie-line power flow $delP_{tie12}$ = 21.5 s > 20 s, $delP_{tie13}$ = 19 s > 18 s, and $delP_{tie23}$ = 18.5 s > 17 s. In addition, the tie-line power flow deviations

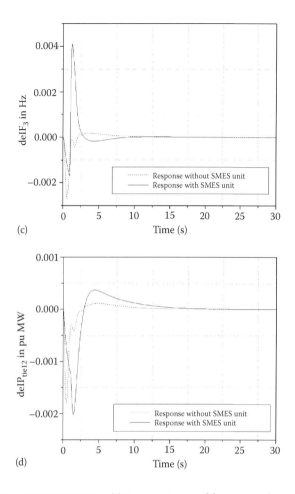

FIGURE 3.4 (CONTINUED)   (c) Comparisons of frequency deviations in area 3. (d) Comparisons of tie-line power flow deviations between areas 1 and 2.   (*Continued*)

between interconnected power systems yield minimum damping oscillations with lesser peak undershoot compared to the response without an SMES unit. Figure 3.4g through i shows the performance comparison of area control error with and without considering an SMES unit in all the areas.

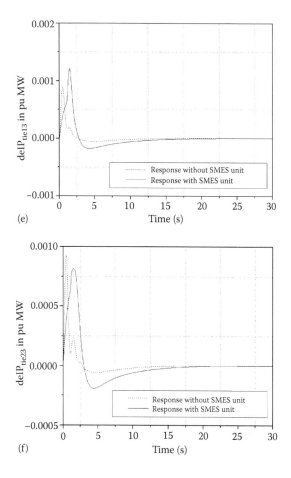

FIGURE 3.4 (CONTINUED) (e) Comparisons of tie-line power flow deviations between areas 1 and 3. (f) Comparisons of tie-line power flow deviations between areas 2 and 3. *(Continued)*

From Figure 3.4g through i it is clearly evident that the area control error is settled quickly compared to the system response without considering an SMES unit. The performance analysis based on the time domain specifications parameters and numerical settling time values are tabulated in Table 3.3 effectively shows

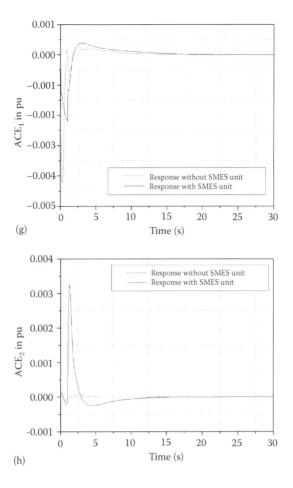

(g)

(h)

FIGURE 3.4 (CONTINUED) (g) Comparisons of area control error deviations in area 1. (h) Comparisons of area control error deviations in area 2.
(*Continued*)

that response with an SMES unit settled quicker compared to the
system response without considering an SMES unit, where $ACE_1 =$
14.5 s > 13.5 s, ACE_2 = 13 s > 12.5 s, ACE_3 = 15.5 s > 14.5 s. In
addition, the area control error of each system yielded minimum
damping oscillations with lesser peak undershoot compared to
the response without an SMES unit.

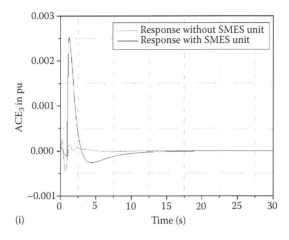

FIGURE 3.4 (CONTINUED)   (i) Comparisons of area control error deviations in area 3.

TABLE 3.3   Comparison of Settling Time Values with and without Considering SMES Unit

Figure Number	Response	Without Considering SMES Unit (s)	With Considering SMES Unit (s)
3.4a	delF1	10.1	10
3.4b	delF2	19	16
3.4c	delF3	16.5	10.1
3.4d	$delP_{tie12}$	21.5	20
3.4e	$delP_{tie13}$	19	18
3.4f	$delP_{tie23}$	18.5	17
3.4g	ACE_1	14.5	13.5
3.4h	ACE_2	13	12.5
3.4i	ACE_3	15.5	14.5

From the simulation result comparisons in the plots and time-domain specifications of the interconnected power system tables it is evident that the system response including an SMES unit settled quicker with lesser damping oscillations compared to the system response without considering an SMES unit in all interconnected areas. Generally, the system response is compared to a system response without considering an SMES unit in the system. The simulation results comparison clearly shows that the system response with an SMES unit improves the system response during sudden load disturbance.

The main objectives that are achieved in this chapter include:

- Developing a three-area interconnected thermal power system with an SMES and without an SMES unit in association with a secondary controller.

- Designing a secondary PID controller for the AGC of an equal three-area interconnected reheat thermal power system, optimizing the controller gain values using the ACO technique.

- Computing the K_p, K_i, and K_d gain values of the PID controller with and without considering an SMES unit.

- Investigating the performances of the AGC of an interconnected reheat thermal system with an SMES energy storage unit.

- Comparing the performance of the AGC reheat thermal system with and without applying the effect of an SMES unit in all areas.

3.6 CONCLUSION

In this work, the AGC of a three-area interconnected thermal power plant is investigated with a PID controller considered as a secondary controller. The gain values of the PID controller are optimized using the ACO method with the ITAE objective function. The controller's gain values are optimized by using the ACO method with and without considering the effect of an SMES unit in all interconnected power plants areas.

The performance of a system with an SMES energy storage unit is compared with performance of a system without considering an energy storage unit. The simulation results show the SMES unit effectively damping out the oscillations and reducing the settling time of the system response (frequency of system, tie-line power flow between interconnected areas, and area control error) during sudden load disturbance. The numerical values and response comparisons are given in a table and plot. It is clearly proven that the system response with an SMES unit settled quickly compared to the system response without considering an SMES unit in all areas.

CHAPTER 4

Flower Pollination Algorithm Optimized PID Controller for Performance Improvement of Multiarea Interconnected Thermal Power System with Nonlinearities

In this chapter, the bio-inspired flower pollination algorithm (FPA) is proposed for optimization of the PID controller gain values in the automatic generation control (AGC) of an interconnected thermal power system. The interconnected power system includes a governor dead band (GBD) and the generation rate constraint (GRC) nonlinearities in all areas. Each area of the interconnected thermal system comprises a governor, reheater, generator, and speed regulator units. The bio-inspired algorithms (BIAs) are implemented to tune the controller gain values in the investigated system to analyze the behavior of the power system during one percent step load perturbation (1% SLP) in area 1. The superiority of the proposed method is proved by comparing the performance of the optimized controller response with the genetic algorithm (GA), simulated annealing (SA), and particle swarm optimization (PSO) techniques in the same analyzed system. Finally, it is clearly evident that the proposed FPA optimized PID controller demonstrated superior performance over the SA, GA, and PSO optimized PID controller during sudden load disturbance.

The chapter is arranged as follows. In the "System Investigated" section, the transfer function and Simulink® model of the investigated power system is presented. The design of the PID controller and structure is given in the "Controller Design and Objective Function" section. The details and tuning of the controller gain using the FPA are introduced. Then an analysis of the results is given. Finally, conclusions about the proposed algorithm-based optimization technique and system performance are drawn.

4.1 SYSTEM INVESTIGATED

In this work, the AGC of a multiarea interconnected power system is considered and represented in the investigated power system with each area including a single-area thermal power generating system. In order to study the dynamic behaviors of the investigated power system, 1% SLP in applied into the thermal area 1 of system by considering GRC and GDB nonlinearities. For controlling frequency deviations of the system and tie-line power flow deviations between connected power systems, a PID controller is equipped in each area as a secondary controller. The MATLAB® Simulink transfer function model of a multiarea interconnected power system with nonlinearity is given in Figure 4.1 [70]. Each area of the investigated power system incorporates a single stage reheat turbine, governor, and reheater unit with GRC and GDB nonlinearity effects in all three interconnected power systems. All three areas are interconnected via a tie-line, and a PID controller is considered as secondary controller for regulating the power system response during sudden load demand conditions.

4.2 CONTROLLER DESIGN AND OBJECTIVE FUNCTION

In this work, a popular feedback PID controller is considered as a secondary controller for regulating the system response during a sudden load disturbance. Fundamentally, it includes three terms: proportional term, integral term, and derivative term. The rise of the system response is reduced with help of the proportional controller, the integral controller eliminates the steady state in their response, and stability of the system is increased by the derivative controller. The controller design plays a crucial role for achieving a better controlled response for the interconnected multiarea interconnected power systems during sudden load demand.

The dynamic response of the controlled response should have minimum settling time with minimal peak overshoot and undershoot in the response. From the literature survey, it is found that many bio-inspired algorithm–based optimization techniques are introduced for getting optimal controller gain values. The flower pollination algorithm is proposed to tune controller gain values by considering the integral square error (ISE) cost function. The literature shows that many researchers have used the ISE cost function for tuning of PID controller gain values in an interconnected power system. For the design of controllers tuned based on BIA optimization techniques, the cost function is first defined

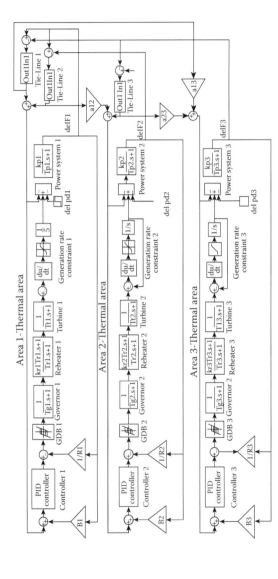

FIGURE 4.1   Transfer function simulink model of a three-area interconnected power system with nonlinearity effects.

based on required specifications and constraints [71]. The expression for the ISE cost function is given as follows:

$$J = ISE = \int_{0}^{T} (\Delta f_{i,j} + \Delta P_{tie\ i,j})^2\ dt \tag{4.1}$$

where J is the performance index, t is the simulation time period, Δf is the frequency deviations, and ΔP_{tie} is the tie-line power deviations (i,j = area 1, 2, 3).

The ACE represents the linear combination of changes in system frequency and tie-line power deviations between interconnected systems. ACE is the input to the proposed controller, and the generated control output signals are u_1, u_2, and u_3. The values of the ACE for each area are given in the following equations:

$$ACE_1 = \Delta P_{tie1} + B_1 \Delta f_1 \tag{4.2}$$

$$ACE_2 = \Delta P_{tie2} + B_2 \Delta f_2 \tag{4.3}$$

$$ACE_3 = \Delta P_{tie3} + B_3 \Delta f_3 \tag{4.4}$$

where ACE is the area control error, ΔP_{tie} is the tie-line power deviation, B is the frequency bias constant, and Δf is the frequency deviation.

The control signal generated by the PID controller for each area is given by

$$u_1 = -K_{p1} ACE_1 - K_{i1} \int ACE_1 - K_{d1} \frac{dACE_1}{dt} \tag{4.5}$$

$$u_2 = -K_{p2} ACE_2 - K_{i2} \int ACE_2 - K_{d2} \frac{dACE_2}{dt} \tag{4.6}$$

$$u_3 = -K_{p3} ACE_1 - K_{i3} \int ACE_3 - K_{d3} \frac{dACE_3}{dt} \tag{4.7}$$

where ACE is the area control error. The maximum/minimum values of the controller parameters are chosen as 1 and 0, respectively.

4.3 FLOWER POLLINATION ALGORITHM (FPA)

Nowadays, the bio-inspired algorithms have a significant role in solving many optimization problems in the field of engineering. The FPA is inspired by the exploit nature of the flower pollination pattern. The FPA was initially developed by Xin-She Yang in the year 2013. In general, 80% of natural plants are flowering plants [72–77] that use their own species for reproduction process. Pollination is defined as exchanging of one flower's pollen to another flower either in another plant or similar plant. The pollen-exchanging process occurs with the help of insects, birds, bats, and other animals. The pollination process is classified into two types: abiotic and biotic. Again it is divided into cross- and self-pollination. The self-pollination process occurs between the same flowering plant. The cross-pollination process occurs between the flowers from the same flowering plants or flowers from the different plants. The process of reproduction and fittest plant species survival is a process of pollination. Based on the behavior of the Lévy weight, few insects and bees fly some distance for obeying Lévy distribution. Flower pollination process is carried out by birds or insects.

The FPA has different pollination process characteristics, based on the flower pollinator behavior and constancy, and are characterized [76] as follows:

1. Cross-pollination and biotic pollination are considered global pollination processes with Lévy flights performance while carrying pollinators.

2. Self-pollination and abiotic are considered local pollination.

3. The tendencies of individual flower constancy of pollinators to visit certain flower plants or morphs within a plant, bypassing the other flower species available.

4. Global pollinations and local pollinations are measured by a switch with p∈ [0, 1] probability. Based on the physical closeness and other parameters like wind, local pollination has a major fraction of p in the overall pollination performance.

The implemented FPA used to get optimal gain values in PID controllers in the three-area interconnected power system is given as follows:

FPA Algorithm

Start
Determine the objective function: min $f(x)$,
Where $x = (kp_d, ki_d, kd_d)$ and $d = 1,2, 3$ (Eq. 1)
Initialize a population of random solution with n flowers
      Find the best solution g_{pidval} in the initial population
Define a switch probability $p∈[0, 1]$
while (t <Max Generation)
for $i = 1: n$(all n flowers in the population)
if r and <p,
Draw a (9-dimensional) step vector L which obeys a Lévy distribution
Global pollination process obtained using $x_i^{t+1} = x_i^t + L\left(g_{pidval} - x_i^t\right)$
else
Draw ε from a uniform distribution in [0,1]
Choose randomly the value of j and k among all the solutions
Do the process of local pollination using $x_i^{t+1} = x_i^t + \varepsilon\left(x_j^t - x_k^t\right)$
end if
Estimate the new solutions
Update solutions: update the population, If new solutions are better,
end for
Find the solution of current best g_{pidval}
end while
End

TABLE 4.1   Parameter Settings of Proposed FPA

Number of flowers =10
Switch probability = 0.8

In this current work, each generated solution is treated as a flower. Each flower is in the form of (Kp_d, Ki_d, Kd_d), within minimum and maximum the value of 0 and 1, respectively, for the three-area interconnected thermal power plant. The ISE cost function is utilized for getting the best solutions from the initial population of flower. The solution regeneration is carried out again as per the standard FPA [73]. The proposed FPA is used for optimization of the PID controller parameter in the AGC of a three-area interconnected thermal power system with GRC and GDB nonlinearity effects. The parameter settings of the proposed optimization algorithm for tuning of the PID controller gain values are given in Table 4.1.

4.4 SIMULATION RESULT AND ANALYSIS

The desirable response and extended stability are crucial for any controlled process, but achieving these constraints simultaneously are more critical. So, there is a need for satisfaction between quick response and stability. These issues are overcome by selecting suitable controller and optimizing controller gain values in the AGC of the multiarea interconnected power system. Consequently, the suitable controller gain values are obtained by using the proposed FPA for the AGC of the multiarea power generating system by the inclusion of nonlinearity effects with ISE cost function.

By considering the ISE cost function, the multiarea interconnected power generating system with nonlinearity effects is simulated with SA, GA, and PSO optimization techniques and the proposed FPA. The SA, GA, PSO, and FPA optimized gain values are depicted in Table 4.2.

TABLE 4.2   Gain Values of PID Controller Optimized by Using GA, PSO, and FPA Techniques

Optimization Technique Controller		SA	GA	PSO	FPA
Proportional gain (K_p)	K_{p1}	0.814724	0.976632	0.959353	0.960225
	K_{p2}	0.913376	0.384582	0.657204	0.51789
	K_{p3}	0.278498	0.980997	0.885267	0.852539
Integral gain (K_i)	K_{i1}	0.905792	0.959141	0.993896	0.999329
	K_{i2}	0.632359	0.233516	0.369237	0.872256
	K_{i3}	0.546882	0.500273	0.334939	0.312738
Derivative gain (K_d)	K_{d1}	0.126987	0.750245	0.807429	0.90084
	K_{d2}	0.09754	0.096227	0.809657	0.939481
	K_{d3}	0.232341	0.570256	0.997222	0.885913

Frequency deviations in area 1, 2, and 3 are given in Figure 4.2a through c during 1% SLP in area 1 and by considering nonlinearity effects in all connected areas. The frequency deviation comparisons in Figure 4.2a through c show that the proposed FPA tuned controller yielded the fastest settled response compared to the other techniques' tuned controller performance with f_1 = 60 s > 26.8 s > 24.5 s > 24 s, f_2 = 60 s > 27 s > 26 s > 14.5 s, and f_3 = 60 s > 27 s > 26 s > 16.5 s. Figure 4.2d through f shows the tie-line power flow deviations between interconnected power system in between areas 1 and 2, areas 1 and 3, and areas 2 and 3, respectively. From these comparisons, it is shown that the proposed FPA tuned controller yielded the fastest settled response compared to the other techniques' tuned controller performance (delP_{tie12} = 60 s > 44 s > 48 s > 37.5 s, delP_{tie13} = 60 s > 60 s > 48 s > 40 s, delP_{tie23} = 60 s > 60 s > 52 s > 30.5 s).

Similarly, the area control error of areas 1, 2, and 3, respectively, is shown in Figure 4.2g through i with the tuned PID controller response of different BIAs in the investigated power system.

Based on the area control error deviations comparisons, it is shown that the proposed optimization technique tuned controller

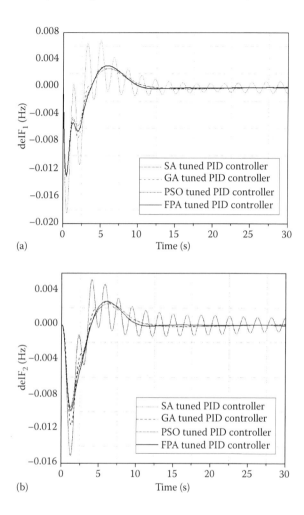

(a)

(b)

FIGURE 4.2 (a) Frequency deviations in area 1 with GDB and GRC nonlinearities. (b) Frequency deviations in area 2 with GDB and GRC nonlinearities. *(Continued)*

effectively gives the better control signal to the power system during sudden load demand. From the response comparisons and the numerical values in the tables, the proposed FPA optimized PID controller yielded better controller response that was smoother and faster compared to the other optimization techniques' tuned

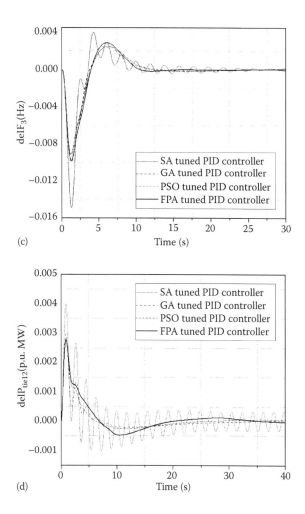

FIGURE 4.2 (CONTINUED)   (c) Frequency deviations in area 3 with GDB and GRC nonlinearities. (d) Tie-line power flow deviations between areas 1 and 2 with GDB and GRC nonlinearities.   (*Continued*)

controller response during sudden load demand in a multiarea interconnected reheat thermal power system by considering nonlinearity effects. Table 4.3 reports the overshoot, undershoot, and the settling time values of frequency deviations and tie-line power flow deviations between control areas of the power system

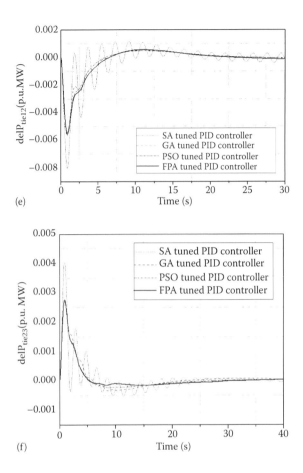

FIGURE 4.2 (CONTINUED) (e) Tie-line power flow deviations between areas 1 and 3 with GDB and GRC nonlinearities. (f) Tie-line power flow deviations between areas 2 and 3 with GDB and GRC nonlinearities.

(*Continued*)

without GRC nonlinearity. It is clearly evident that the proposed FPA technique tuned controller settled more quickly compared to the tuned controller performance of the other techniques. In addition, SA-PID controller response yield minimum peak undershoot and overshoot compared to the tuned controller response of the other optimization techniques.

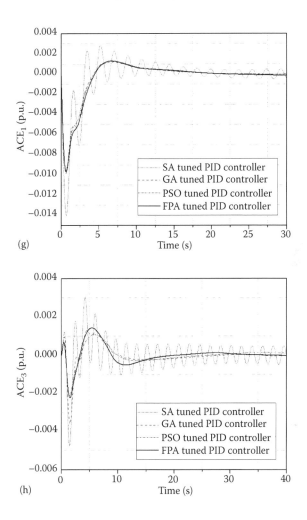

(g)

(h)

FIGURE 4.2 (CONTINUED)   (g) Area control error in area 1 with GDB and GRC nonlinearities. (h) Area control error in area 2 with GDB and GRC nonlinearities.   (*Continued*)

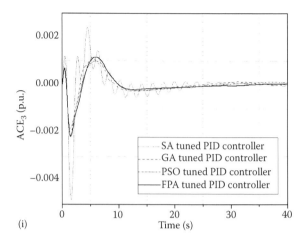

FIGURE 4.2 (CONTINUED) (i) Area control error in area 3 with GDB and GRC nonlinearities.

This work proposed an efficient bio-inspired algorithm–based optimization technique for tuning of controller gain values to solve the AGC issue in a three-area interconnected thermal power system during sudden load demand by considering the nonlinearity effects in the system. The major motivation of the chapter can be specified as follows:

1. The AGC of the three-area interconnected thermal power plant by the inclusion of nonlinearity effects.

2. The dynamic performance of the three-area interconnected thermal system is examined by applying 1% SLP in area 1 and considering the GRC and GDB nonlinearity effects.

3. Secondary PID controller gain values are optimized for regulating system performance during emergency conditions.

4. The controller gain values are tuned and optimized by using different bio-inspired algorithm–based optimization techniques. Performances of the tuned controllers of each optimization technique are compared to find the supremacy technique.

TABLE 4.3   Overshoot, Undershoot, and Settling Time of Frequency Deviations and Tie-Line Power Flow between Control Areas of Power System without GRC Nonlinearity

		Parameter		
Response	Technique	Peak Overshoot	Peak Undershoot	Settling Time (s)
Δf_1 (Hz)	SA	0.5×10^{-3}	-0.0184	60
	GA	2.9×10^{-3}	-0.0132	26.8
	PSO	2.7×10^{-3}	-0.0128	24.5
	FPA	$\mathbf{3.2 \times 10^{-3}}$	$\mathbf{-0.0128}$	**24**
Δf_2 (Hz)	SA	5.3×10^{-3}	-0.015	60
	GA	1.8×10^{-3}	-0.0116	27
	PSO	3×10^{-3}	-9.5×10^{-3}	26
	FPA	$\mathbf{2.75 \times 10^{-3}}$	$\mathbf{-9.8 \times 10^{-3}}$	**14.5**
Δf_3 (Hz)	SA	4×10^{-3}	-0.0148	60
	GA	2.5×10^{-3}	-10×10^{-3}	27
	PSO	3×10^{-3}	-9.2×10^{-3}	26
	FPA	$\mathbf{3.4 \times 10^{-3}}$	$\mathbf{-9.8 \times 10^{-3}}$	**16.5**
ΔP_{tie1} (p.u. MW)	SA	0.3×10^{-3}	-8×10^{-4}	60
	GA	-5×10^{-4}	-5.1×10^{-3}	44
	PSO	4.9×10^{-4}	5.442×10^{-3}	48
	FPA	$\mathbf{6 \times 10^{-4}}$	$\mathbf{5.6 \times 10^{-3}}$	**37.5**
ΔP_{tie2} (p.u. MW)	SA	4×10^{-3}	-7×10^{-4}	60
	GA	2.88×10^{-3}	-1.2×10^{-4}	60
	PSO	2.75×10^{-3}	2.4×10^{-4}	48
	FPA	$\mathbf{2.85 \times 10^{-3}}$	$\mathbf{4.8 \times 10^{-4}}$	**40**

(Continued)

TABLE 4.3 (CONTINUED)   Overshoot, Undershoot, and Settling Time of
Frequency Deviations and Tie-Line Power Flow between Control Areas of
Power System without GRC Nonlinearity

Response	Technique	Parameter		
		Peak Overshoot	**Peak Undershoot**	**Settling Time (s)**
ΔP_{tie3} **(p.u. MW)**	SA	4.1×10^{-3}	-3.6×10^{-4}	60
	GA	2.7×10^{-3}	1.8×10^{-4}	60
	PSO	2.7×10^{-3}	2.4×10^{-4}	52
	FPA	$\mathbf{2.75 \times 10^{-3}}$	$\mathbf{1.6 \times 10^{-4}}$	**30.5**

4.5 CONCLUSION

The recently developed bio-inspired flower pollination algorithm
is proposed to optimize the PID controller gain values to analyze the performance of the AGC of a multiarea interconnected
thermal power plant with GRC and GDB nonlinearity effects. The
dynamic behavior of the proposed FPA optimized controller performance is compared with the tuned controller response of GA,
SA, and PSO techniques on the same system by applying 1% SLP
in area 1.

The combined response of the system shows that the proposed
algorithm tuned controller gives a better controlled response
in terms of minimum settling time compared to the other optimization techniques. However, SA-PID controller responses
having minimum undershoot and overshoot value compared to
the tuned controller response of other optimization techniques
(GA and PSO).

CHAPTER 5

Challenges and Future Perspectives

T HE NUMEROUS FASCINATING ISSUES for further research related to load frequency control and automatic generation control of interconnected Single area and multiarea power-generating systems magnetize several researchers. The pattern of load disturbance is replaced from the step load disturbance into continuous load demand for a specific time interval in the system. This book presents several studies of different power plant interconnections and increases in number of interconnections with other power-generating systems, such as wind power systems, gas turbine power systems, and solar photovoltaic (PV) cells.

To improve the dynamic performance of interconnected power systems, besides the traditional bio-inspired algorithms [78–107] are being developed for tuning the secondary controller gain values for better controlled performance during sudden load disturbance and continuously varying load conditions. The thermal power system is equipped with a single-stage reheat turbine to provide steam for additional usage. In future work, single-stage reheat turbines can be replaced with two, three, or multistage reheat turbines. The power system needs a more efficient secondary controller for generating good quality power supply to consumers. However, it will also be relatively useful for practical applications and operations of power systems during an emergency load demand situation.

Future research work can be made for regulation of wind energy conversion systems with the support of electronic control of the induction generator and synchronous drives. This study can be further applied to interconnections of different renewable energy sources, such as wind–PV–diesel and wind–biogas–diesel power generating units.

References

1. B. K. Sahu, S. Pati, P. K. Mohanty, S. Panda, "Teaching-learning based optimization algorithm based fuzzy-PID controller for automatic generation control of multi-area power system," *Applied Soft Computing*, 27, 240–249, 2015.

2. R. K. Sahu, S. Panda, T. S. Gorripotu, "A novel hybrid PSO-PS optimized fuzzy PI controller for AGC in multi area interconnected power systems," *Electrical Power and Energy Systems*, 64, 880–893, 2015.

3. J. Nanda, M. Sreedhar, A. Dasgupta, "A new technique in hydrothermal interconnected automatic generation control system by minority charge carrier inspired algorithm," *Electrical Power and Energy Systems*, 68, 259–268, 2015.

4. M. R. Sathya, M. Mohamed Thameem Ansari, "Load frequency control using bat inspired algorithm based dual mode gain scheduling of PI controller for interconnected power system," *Electrical Power and Energy Systems*, 64, 365–374, 2015.

5. I. Pan, S. Das, "Fractional-order load frequency control of interconnected power systems using chaotic multi-objective optimization," *Applied Soft Computing*, 29, 328–344, 2015.

6. P. Dash, L. C. Saikia, N. Sinha, "Comparison of performance of several FACTS devices using Cuckoo search algorithm optimized 2DOF controllers in multi-area AGC," *Electrical Power and Energy Systems*, 65, 316–324, 2015.

7. P. Dash, L. C. Saikia, N. Sinha, "Automatic generation control of multi area thermal system using bat algorithm optimized PD-PID cascade controller," *Electrical Power and Energy Systems*, 68, 364–372, 2015.

8. R. K. Sahu, S. Panda, S. Padhan, "A hybrid firefly algorithm and pattern search technique for automatic generation control of multi area power systems," *Electrical Power and Energy Systems*," 64, 9–23, 2015.

9. B. K. Sahu, T. K. Pati, J. R. Nayak, S. Panda, S. K. Kar, "A novel hybrid LUS-TLBO optimized fuzzy-PID controller for load frequency control of multi-source power system," *Electrical Power and Energy Systems*, 24, 58–69, 2016.

10. G. Shankar, V. Mukherjee V, "Quasi oppositional harmony search algorithm based controller tuning for load frequency control of multi-source multi-area power system," *Electrical Power and Energy Systems*, 75, 289–302, 2016.

11. R. Thirunavukarasu, I. A. Chidamdaram, "PI^2 controller based coordinated control with redox flow battery and unified power flow controller for improved restoration indices in a deregulated power system," *Ain Shams Engineering Journal*, 7(4), 1011–1027, 2016.

12. P. C. Padhan, R. K. Sahu, S. Panda, "Firefly algorithm optimized fuzzy PID controller for AGC of multi-area multi-source power systems with UPFC and SMES," *Engineering Science and Technology, An International Journal*, 19(1), 338–354, 2016.

13. Y. Sharma, L. C. Saukia, "Automatic generation control of a multi-area ST-Thermal power system using Grey Wolf Optimizer algorithm based classical controllers," *Electrical Power and Energy Systems*, 73, 853–862, 2015.

14. R. Francis, I. A. Chidambaram, "Optimized PI+ load-frequency controller using BWNN approach for an interconnected reheat power system with RFB and hydrogen electrolyser units," *Electrical Power and Energy Systems*, 67, 381–392, 2015.

15. M. Shivaie, M. G. Kazemi, M. T. Ameli, "A modified harmony search algorithm for solving load-frequency control of non-linear interconnected hydrothermal power systems," *Sustainable Energy Technologies and Assessments*, 10, 53–62, 2015.

16. S. Padhan, R. K. Sahu, S. Panda, "Application of firefly algorithm for load frequency control of multi-area interconnected power system," *Electric Power Components and Systems*, 42(13), 1419–1430, 2014.

17. S. Prakash, S. K. Sinha, "Simulation based neuro-fuzzy hybrid intelligent PI control approach in four-area load frequency control of interconnected power system," *Applied Soft Computing*, 23, 152–164, 2014.

18. R. Farhangi, M. Boroushaki, S. H. Hosseini, "Load-frequency control of interconnected power system using emotional learning-based intelligent controller," *Electrical Power and Energy Systems*, 36, 76–83, 2012.

19. S. Farook, P. S. Raju, "Feasible AGC controllers to optimize LFC regulation in deregulated power system using evolutionary hybrid genetic firefly algorithm," *Journal of Electrical Systems*, 8(4), 459–471, 2012.

20. M. H. Khooban, T. Niknam, "A new intelligent online fuzzy tuning approach for multi-area load frequency control: Self Adaptive Modified Bat Algorithm," *Electrical Power and Energy Systems*, 71, 254–261, 2015.

21. R. Roy, P. Bhatt, S. P. Ghosal, "Evolutionary computation based three-area automatic generation control," *Expert Systems With Applications*, 37, 5913–5924, 2010.

22. S. A. Taher, M. H. Fini, S. F. Aliabadi, "Fractional order PID controller design for LFC in electric power systems using imperialist competitive algorithm," *Ain Shams Engineering Journal*, 5, 121–135, 2014.

23. B. Anand, A. Ebenezer Jeyakumar, "Load frequency control with fuzzy logic controller considering non-linearities and boiler dynamics," *ACSE*, 8, 15–20, 2009.

24. R. Arivoli, I. A. Chidambaram, "Design of genetic algorithm (GA) based controller for load-frequency control of power systems interconnected with AC-DC TIE-LINE," *International Journal of Computer Science & Emerging Technologies*, 2, 280–286, 2011.

25. A. Demiroren, A. L. Zeynelgil, N. S. Sengor, "The application of ANN technique to load frequency control for three-area power systems," *IEEE Porto Power Tech Proceedings*, Porto, Portugal, September 10–13, 2001.

26. H. Shayeghi, H. A. Shayanfor, "Application of ANN technique based–synthesis to load frequency control of interconnected power and energy systems," *Electrical Power and Energy Systems*, 28, 503–511, 2006.

27. D. Das, M. L. Kothari, D. P. Kothari, P. J. Nanda, "Variable structure control strategy to automatic generation control of interconnected reheat thermal system," *IEE Proceedings D, Control Theory and Applications*, 138(6), 579–585, 1991.

28. S. C. Tripathy, G. S. Hope, O. P. Malik, "Optimization of load-frequency control parameters for power systems with reheat steam turbines and governor dead band nonlinearity," *IEE Proceedings C—Generation, Transmission and Distribution*, 129(1), 10–16, 1982.

29. S. C. Tripathy, T. S. Bhatti, C. S. Jha, O. P. Malik, G. S. Hope, "Sampled data automatic generation control analysis with reheat steam turbines and governor dead-band effects," *IEE Transactions on Power Apparatus and Systems*, 103(5), 1045–1051, 1984.

30. C. T. Pan, C. M. Liaw, "An adaptive controller for power system load-frequency control," *IEEE Transactions on Power Systems*, 4(1), 122–128, 1989.

31. J. Nanda, B. L. Kaul, "Automatic generation control of an interconnected power system," *Proceedings of the Institution of Electrical Engineers*, 125(5), 385–390, 1978.

32. M. L. Kothari, J. Nanda, "Application of optimal control strategy to automatic generation control of a hydrothermal system," *IEE Proceedings D—Control Theory and Applications*, 135(4), 268–274, 1988.

33. Y.-H. Moon, H.-S. Ryu, B. Kim, K.-B. Song, "Optimal tracking approach to load frequency control in power systems," *IEEE Power Engineering Society Winter Meeting*, 1371–1376, 2000.

34. M. Aldeen, J. F. Marsh, "Decentralised proportional-plus-integral design method for interconnected power systems," *IEE Proceedings C—Generation, Transmission and Distribution*, 138(4), 263–274, 1991.

35. M. Omar, M. Solimn, A. M. Abdelghany, F. Bendary, "Optimal tuning of PID controllers for hydrothermal load frequency control using ant colony optimization," *International Journal on Electrical Engineering and Informatics*, 5, 348–356, 2013.

36. Y.-T. Hsiao, C.-L. Chuang, C.-C. Chien, "Ant colony optimization for designing of PID controllers," *IEEE International Symposium on Computer Aided Control Systems Design*, 321–326, 2004.

37. K. Jagatheesan, B. Anand, "Automatic generation control of three area hydro-thermal power systems considering electric and mechanical governor with conventional and ant colony optimization technique," *Advances in Natural and Applied Science*, 8(20), 25–33, 2014.

38. L. C. Saikia, N. Sinha, J. Nanda, "Maiden application of bacterial foraging based fuzzy IDD controller in AGC of a multi-area hydrothermal system," *Electrical Power and Energy Systems*, 45, 98–106, 2013.

39. E. S. Ali, S. M. Abd-Elazim, "Bacteria foraging optimization algorithm based load frequency controller for interconnected power system," *Electrical Power and Energy Systems*, 33, 633–688, 2011.

40. H. Gozde, M. Cengiz Taplamacioglu, I. Kocaarslan, "Comparative performance analysis of artificial bee colony algorithm in automatic generation control for interconnected reheat thermal power system," *Electrical Power and Energy Systems*, 42, 167–178, 2012.

41. M. A. Ebrahim, H. E. Mostafa, S. A. Gawish and F. M. Bendary, "Design of decentralized load frequency based-PID controller using stochastic particle swarm optimization technique," *International Conference on Electric Power and Energy Conversion System*, 1–6, 2009.

42. K. Jagatheesan, B. Anand, M. A. Ebrahim, "Stochastic particle swarm optimization for tuning of PID Controller in load frequency control of single area reheat thermal power system," *International Journal of Electrical and Power Engineering*, 8(2), 33–40, 2014.

43. E. S. Ali, S. M. Abd-Elazim, "Bacteria foraging optimization algorithm based load frequency controller for interconnected power system," *Electrical Power and Energy Systems*, 33, 633–638, 2011.

44. L. C. Saikia, N. Sinha, J. Nanda, "Maiden application of bacterial foraging based fuzzy IDD controller in AGC of a multi-area hydrothermal system," *Electrical Power and Energy Systems*, 45, 98–106, 2013.

45. P. Dash, L. C. Saikia, N. Sinha, "Automatic generation control of multi area thermal system using Bat algorithm optimized PD-PID cascade controller," *Electrical Power and Energy Systems*, 68, 364–372, 2015.

46. R. Francis, I. A. Chidambaram, "Optimized PI+ load-frequency controller using BWNN approach for an interconnected reheat power system with RFB and hydrogen electrolyser units," *Electrical Power and Energy Systems*, 67, 381–392, 2015.

47. P. Dash, L. C. Saikia, N. Sinha, "Comparison of performance of several FACTS devices using Cuckoo search algorithm optimized 2DOF controllers in multi-area AGC," *Electrical Power and Energy Systems*, 65, 316–324, 2015.

48. P. Dash, L. C. Saikia, N. Sinha, "Comparison of performances of several Cuckoo search algorithm based 2DOF controllers in AGC of multi-area thermal system," *Electrical Power and Energy Systems*, 55, 429–436, 2014.

49. R. MacArthur, E. Wilson, *The Theory of Biogeography*, Princeton, NJ, Princeton University Press, 1967.

50. D. Simon, "Biogeography-based optimization," *IEEE Transactions on Evolutionary Computation*, 12(6), 702–713.

51. M. Dorigo, T. Stutzle, *Ant Colony Optimization*, Cambridge, MA, MIT Press, 2004.

52. M. Dorigo, L. Gambardella, M. Middendorf, T. Stutzle, eds., "Special section on ant colony optimization," *IEEE Transactions on Evolutionary Computation*, 6(4), 317–365, 2002.

53. K. Price, R. Storn, "Differential evolution," *Dr. Dobb's Journal*, 22, 18–20, 22, 24, 78, 1997.

54. R. Storn, "System design by constraint adaptation and differential evolution," *IEEE Transactions on Evolutionary Computation*, 3, 22–34, 1999.

55. R. Eberhart, Y. Shi, "Special issue on particle swarm optimization," *IEEE Transactions on Evolutionary Computation*, 8(3), 201–228, 2004.

56. S. C. Tripathy, R. Balasubramaniam P. S. Chandramohan Nair, "Adaptive automatic generation control with SMES in power systems," *IEEE Transactions on Energy Conversion*, 7, 434–441, 1992.

57. K. Tam, P. Kumar, "Application of SMES in an asynchronous link between power systems," *IEEE Transactions on Energy* Conversion, 5, 436–444, 1990.

58. H. A. Peterson, N. Mohan, R. W. Boom, "Superconductive energy storage inductor-converter units for power systems," *IEEE Transactions on Power Apparatus and Systems*, 94, 1337–1348, 1975.

59. Y. Mitani, K. Tsuji, Y. Murkami, "Application of superconducting magnet energy storage to improve power system dynamic performance," *IEEE Transactions on Power Systems*, 3, 1418–1425, 1988.

60. S. Banerjee, J. K. Chatterjee, S. C. Tripathy, "Application of magnetic energy storage unit as load frequency stabilizer," *IEEE Transactions on Energy Conversion*, 5, 46–51, 1990.

61. K. S. Tam, P. Kumar, "Applications of superconductive magnetic energy storage in an asynchronous link between power systems," *IEEE Transactions on Energy Conversion*, 5, 436–444,1990.

62. S. C. Tripathy, R. Balasubramaniam, P.S. Chandramohan Nair, "Effect of superconducting magnetic energy storage on automatic generation control considering governor dead and boiler dynamics," paper presented at the IEEE PES International Power Meeting, India, 1990, (Paper No. 90 IC 588-4 PWRS).

63. S. Pothiya, I. Ngamroo, W. Kongprawechnon, "Design of optimal fuzzy logic-based PID controller using multiple tabu search algorithm for AGC including SMES units," *8th International Power Engineering Conference (IPEC 2007)*, 838–843, 2007.

64. R. J. Abraham, D. Das, A. Patra, "Damping oscillations in tie-power and area frequencies in a thermal power system with SMES-TCPS combination," *Journal of Electrical Systems*, 1(1), 71–80, 2011.

65. A. Roy, S. Dutta, P. K. Roy, "Automatic generation control by SMES-SMES controllers of two-area hydro-hydro system," *Proceedings of 2014 1st International Conference on Non Conventional Energy (ICONCE 2014)*, 302–307, 2014.

66. S. Padhan, R. K. Sahu, S. Pandhan, "Automatic generation control with thyristor controlled series compensator including superconducting magnetic energy storage units," *Ain Shams Engineering Journal*, 5, 759–774, 2014.

67. P. Bhatt, S. P. Ghoshal, R. Roy, "Load frequency stabilization by coordinated control of thyristor controlled phase shifters and superconducting magnetic energy storage for three types if interconnected two-area power systems," *Electrical Power and Energy Systems*, 32, 1111–1124, 2010.

68. S. Chaine, M. Tripathy, "Design of an optimal SMES for automatic generation control of two-area thermal power system using Cuckoo search algorithm," *Journal of Electrical Systems and Information Technology*, 2(1), 1–13, 2015.

69. K. Jagatheesan, B. Anand, "Dynamic performance of multi-area hydro thermal power systems with integral controller considering various performance indices methods," *Proceedings of the IEEE International Conference of Emerging Trends in Science, Engineering and Technology (INCOSET)*, 474–478, 2012.

70. B. Anand, A. E. Jeyakumar, "Fuzzy logic based load frequency control of hydro-thermal system with non-linearities," *International Journal of Electrical and Power Engineering*, 3(2), 112–118, 2009.

71. K. Jagatheesan, B. Anand, N. Dey, M. Omar, V. E. Balas, "AGC of multi-area interconnected power systems by considering different cost functions and Ant Colony Optimization technique based PID controller," *An International Journal Intelligent Decision Technologies* (accepted for publication), 2016.

72. P. Willmer, *Pollination and Floral Ecology*, Princeton, NJ, Princeton University Press, 2011.

73. X.-S. Yang, "Flower pollination algorithm for global optimization," arXiv:1312.5673v1, 2013.

74. B. J. Glover, "Understanding flowers and flowering: An integrated approach," *Annals of Botany*, 103(1), vi–vii, 2009.

75. N. M. Waser, "Flower constancy: Definition, cause and measurement," *The American Naturalist*, 127(5), 596–603, 1986.

76. I. Pavlyukevich, "Lévy flights, non-local search and simulated annealing," *Journal of Computational Physics*, 226, 1830–1844, 2007.

77. X. S. Yang, S. Deb, M. Loomes, M. Karamanoglu, "A framework for self-tuning optimization algorithm," *Neural Computing and Applications*, 23(7–8), 2051–2057, 2013.

78. K. Jagatheesan, B. Anand, K. Baskaran, N. Dey, A. S. Ashour, V. E. Bala, "Effect of non-linearity and boiler dynamics in automatic generation control of multi-area thermal power system with proportional-integral-derivative and ant colony optimization technique," In *Recent Advances in Nonlinear Dynamics and Synchronization*, Studies in Computational Intelligence 109, K. Kyamakya, W. Mathis, R. Stoop, J. Chedjou, Z. Li, eds., 89–110, Springer, 2018.

79. K. Jagatheesan, B. Anand, S. Samanta, N. Dey, V. Santhi, A. S. Ashour, V. E. Balas, "Application of flower pollination algorithm in load frequency control of multi-area interconnected power system with non-linearity," *Neural Computing and Applications (NCAA)*, 28, 475–488, 2016.

80. J. Kaliannan, B. Anand, N. Dey, A. S. Ashour, S. C. Satapathy, "Performance evaluation of objective functions in automatic generation control of thermal power system using ant colony optimization technique designed proportional-integral-derivative controller," *Electrical Engineering*, 1–17, 2017.

81. N. Dey, A. S. Ashour, S. Beagum, D. S. Pistola, M. Gospodinov, E. P. Gospodinova, J. M. R. S. Tavares, "Parameter optimization for local polynomial approximation based intersection confidence interval filter using genetic algorithm: An application for brain MRI image de-noising," *Journal of Imaging*, 1, 60–84, 2015.

82. J. Kaliannan, A. Baskaran, S. Samanta, N. Dey, V. E. Balas, "Particle swarm optimization based parameters optimization of PID controller for load frequency control of multi-area reheat thermal power systems," *International Journal of Advanced Intelligence Paradigms*, 9, 464–489, 2017.

83. J. Kaliannan, A. Baskaran, N. Dey, A. S. Ashour, "Ant Colony Optimization algorithm based PID controller for LFC of single area power system with non-linearity and boiler dynamics," *World Journal of Modelling and Simulation*, 12, 3–14, 2016.

84. A. S. Ashour, S. Samanta, N. Dey, V. E. Balas, J. Kaliannan, A. Baskaran, "Design of proportional-integral-derivative controller for AGC of multi-area power thermal systems using firefly algorithm," *IEEE/CAA Journal of Automatica Sinica*, (99), 1–14.

85. S. Samanta, N. Dey, P. Das, S. Acharjee, S. S. Chaudhuri, "Multilevel threshold based gray scale image segmentation using cuckoo search," Proceedings of *International Conference on Emerging Trends in Electrical, Communication And Information Technologies (ICECIT)*, 27–34, 2012.

86. K. Jagatheesan, B. Anand, N. Dey, M. Omar, V. E. Balas, "AGC of multi-area interconnected power systems by considering different cost functions and Ant Colony Optimization technique based PID controller," *Intelligent Decision Technologies*, 11(1), 29–38, 2017.

87. J. Kaliannan, B. Anand, N. Dey, A. S. Ashour, V. E. Balas, "Load frequency control of multi-area interconnected thermal power system: Artificial intelligence based approach," *International Journal of Automation and Control*, 12(1), 126–152, 2016.

88. K. Jagatheesan, B. Anand, N. Dey, "Automatic generation control of thermal-thermal-hydro power systems with PID controller using ant colony optimization," *International Journal of Service Science, Management, Engineering, and Technology (IJSSMET)*, 6(2), 18–34.

89. K. Jagatheesan, B. Anand, N. Dey, A. S. Ashour, "Artificial intelligence in performance analysis of load frequency control in thermal-wind-hydro power systems," *International Journal of Advanced Computer Science and Applications*, 6(7), 203–212, 2015.

90. J. Kaliannan, A. Baskaran, N. Dey, T. Gaber, A. E. Hassanien, "A design of PI controller using stochastic particle swarm optimization in load frequency control of thermal power systems," paper presented at 5th International Conference on Circuits, Control, Communication, Electricity, Electronics, Energy, System, Signal and Simulation (CES-CUBE 2015), at Pattaya, Thailan, 2015d.

91. S. Chakraborty, S. Samanta, A. Mukherjee, N. Dey, S. S. Chaudhuri, "Particle swarm optimization based parameter optimization technique in medical information hiding," paper presented at 2013 IEEE International Conference on Computational Intelligence and Computing Research (ICCIC), Madurai, India, December 26–28, 2013.

92. K. Jahatheesan, B. Anand, N. Dey, M. A. Ebrahim, "Design of proportional-integral-derivative controller using Stochastic Particle Swarm Optimization technique for single area AGC including SMES and RFB units," paper presented at International Conference on Computer and Communication Technologies, Hyderabad, India, July 24–26, 2015.

93. S. Chatterjee, S. Sarkar, S. Hore, N. Dey, A. S. Ashour, V. E. Balas, "Particle swarm optimization trained neural network for structural failure prediction of multi-storied RC buildings," *Neural Computing and Applications*, 28(8), 2005–2016, 2017.

94. N. Dey, S. Samanta, X.-S. Yang, S. S. Chaudhri, A. Das, "Optimisation of scaling factors in electrocardiogram signal watermarking using cuckoo search," International *Journal of Bio-Inspired Computation (IJBIC)*, 5(5), 315–326, 2013.

95. W. B. Ab. Karaa, A. S. Ashour, D. B. Sassi, P. Roy, N. Kausar, N. Dey, "MEDLINE text mining: An enhancement genetic algorithm based approach for document clustering," In *Applications of Intelligent Optimization in Biology and Medicine: Current Trends and Open Problems*, A. E. Hassanien, C. Grosan, M. Fahmy Tolba, eds., 267–287, Springer, 2016.

96. K. Jagatheesan, B. Anand, N. Dey, A. S. Ashour, V. E. Balas, "Load frequency control of hydro-hydro system with fuzzy logic controller considering non-linearity," paper presnted at World Conference on Soft Computing, Berkeley, California, May 22–25, 2016.

97. M. Chennoufi, F. Bendella, M. Bouzid, "Multi-agent simulation collision avoidance of complex system: Application to evacuation crowd behavior," *International Journal of Ambient Computing and Intelligence (IJACI)*, 9(1), 43–59, 2018.

98. S. Chakraborty, N. Dey, S. Samanta, A. S. Ashour, V. E. Balas, "Firefly algorithm for optimized non-rigid demons registration," in *Bio-Inspired Computation and Applications in Image Processing*, X. S. Yang and J. P. Papa, eds., 221–237, Elsevier, 2016.

99. D. Juneja, A. Singh, R. Singh, S. Mukherjee, A thorough insight into theoretical and practical developments in multiagent system, *International Journal of Ambient Computing and Intelligence (IJACI)*, 8(1), 23–49, 2017.

100. S. Samanta, A. Choudhury, N. Dey, A. S. Ashour, V. E. Balas, "Quantum inspired evolutionary algorithm for scaling factors optimization during manifold medical information embedding," in *Quantum Inspired Computational Intelligence: Research and Applications*, S. Bhattacharyya, U. Maulik, P. Dutta, eds., 285–326, Morgan Kaufmann, 2017.

101. D. Acharjya, A. Anitha, "A comparative study of statistical and rough computing models in predictive data analysis, *International Journal of Ambient Computing and Intelligence (IJACI)*, 8(2), 32–51, 2017.

102. G. N. Nguyen, K. Jagatheesan, A. S. Ashour, B. Anand, N. Dey, "Ant colony optimization based load frequency control of multi-area interconnected thermal power system with governor deadband nonlinearity," paper presented at World Conference on Smart Trends in Systems, Security and Sustainability, London, February 15–16, 2017.

103. H. W. Guesgen, S. Marsland, "Using contextual information for recognising human behaviour" *International Journal of Ambient Computing and Intelligence (IJACI)*, 7(1), 27–44, 2016.

104. N. Dey, S. Samanta, S. Chakraborty, A. Das, S. S. Chaudhuri, J. S. Suri, "Firefly algorithm for optimization of scaling factors during embedding of manifold medical information: An application in ophthalmology imaging," *Journal of Medical Imaging and Health Informatics*, 4(3), 384–394, 2014.

105. S. Chatterjee, N. Dey, A. S. Ashour, C. V. A. Drugarin, "Electrical energy output prediction using cuckoo search supported artificial neural network," paper presented at World Conference on Smart Trends in Systems, Security and Sustainability, London, February 15–16, 2017.

106. N. A. Mhetre, A. V. Deshpande, P. N. Mahalle, "Trust management model based on fuzzy approach for ubiquitous computing," *International Journal of Ambient Computing and Intelligence (IJACI)*, 7(2), 33–46, 2016.

107. R. Kumar, A. Rajan, F. A. Talukdar, N. Dey, V Santhi, V. E. Balas, "Optimization of 5.5 GHz CMOS LNA parameters using firefly algorithm," *Neural Computing and Applications (NCAA)*, 28(12), 3765–3779, 2017.

Index

Printed and bound by CPI Group (UK) Ltd, Croydon, CR0 4YY

24/10/2024

01778282-0002